WHY THEORY

Finally, a Unified Theory Everyone Can Understand and Albert Einstein Would Love!

Authored By

Robert Kaufman

ISBN-10: 1533329915
ISBN-13: 978-1533329912

Finally, an entertaining, easy to read, explanation of the great mysteries and how the universe really works.

Shocking answers to the world's most baffling questions

"WHY THEORY"

If you don't know why, you don't know anything.

Einstein was right... Niels Bohr was wrong... after all.

Be Smarter than Isaac Newton, Niels Bohr, Nicola Tesla, Albert Einstein, Stephen Hawking and everyone you know

"WHY THEORY"

Table of Contents

Introduction

During the 1990s I lived in Cold Spring Harbor, Long Island, New York, home of the world famous Cold Spring Harbor Laboratory.

My wife and I entertained a group of Russian born scientists who were teaching, studying and working locally. I quickly realized that I was the ninth smartest person in the room- there were eight of them and me.

Our discussions bounced from subject to subject- it was a feast, a banquet of knowledge. I had become a businessman by necessity but I sold my company and was able to devote attention to my true avocation- science.

I peppered them with questions on subjects from the Big Bang Theory to nuclear energy. (I had recently read the 1992 book "The Big Bang Never Happened" by Eric Lerner and several other books on science). We talked about Newton, Einstein, Niels Bohr, the uncertainty principle, matter and anti-matter.

Now everyone knows from old Star Trek episodes that matter and anti-matter destroy each other releasing more energy than a nuclear explosion. I asked why don't we use this reaction to supply the world's energy needs. They explained that anti-matter does not naturally exist in our universe. Anti-matter can be created in a laboratory but it takes enormous amounts of energy to create even the smallest amount of anti-matter. I could see they were amused by my naïveté.

I asked them why does matter destroy anti-matter. Their answer surprised me and maybe it will surprise you. "We don't know". It did not seem like a hard question so I must have looked surprised. Again they were amused and added- We don't know the answer to most WHY questions.

Why does gravity work? We don't know. E=MC2 means energy can be converted to matter and matter can be converted to energy but we don't know why. We just don't know. We don't even know why magnetism works? We know how it works but not why it works.

Magnetism? Children play with magnets. Moms put magnets on refrigerators. How is it possible that we can send rockets to Mars but we cannot explain magnetism? Why do magnets have poles? Why do like poles repel each other? Why do magnets create lines of force?

I decided to try my hand at developing a workable theory of magnetism or at least some hypothesis that would provide answers to those WHY questions. What developed was more than a theory of magnetism, it was a theory of everything. Why does matter destroy anti-matter? Why does gravity exist? Why does energy become matter or matter become energy? Why have we not been able to explain one of the most baffling mysteries of the universe- the wave/particle duality?

After years of scientific discussions, research, study, fact checking, updating, and procrastination my sons have prevailed upon me to write this book.

"Why Theory" will make you smarter than everyone you know. Now you can be the smartest person in the room.

"Why Theory" Five Friends

News reporters are very familiar with this slant rhyme: "I have five friends who taught me everything I know: They are who, what, where, when, and how".

Scientists also have five friends but they are a little different. I have five friends who taught me everything I know: They are what, where, when, how and WHY.

There is no real understanding in science without understanding WHY. You have a pot of water over a flame. The water disappears. WHY? Maybe heat destroys water. Maybe the metal pot absorbs water when heated. Maybe the water leaves our universe through a wormhole into another universe. Maybe the water just decides to go out of existence. We know all these explanations are silly because scientific experiments have proven that the liquid water becomes water vapor, a gas. The water that was in the pot is now in the air. Mystery solved. Not quite. WHY does a liquid become a gas?

The scientific answer is that water is composed of 2 hydrogen atoms and one oxygen atom and 10 electrons. The three atoms and 10 electrons form one molecule of water. The electrons orbit around the three nuclei. The extra energy from the flame causes the electrons to expand their orbit creating more distance between the molecules of water. This expansion changes the liquid water to water gas (vapor) which goes into the air. Now the mystery is solved. Not quite.

WHY does the energy from the flame cause the electron orbits to expand? The answer to that basic question has been missing. That is one of the many questions **"Why Theory"** will attempt to answer later in this book.

When the WHY answer is missing or when an incorrect WHY answer is accepted as truth, scientific progress is stymied. Brilliant people will waste their intellects and efforts, sometimes a lifetime, going in the wrong direction.

The moon orbits the earth. The sun appears to orbit the earth. The best minds assumed the earth was the center of the universe. This geocentric model had problems of course. When the planets were not orbiting the earth "correctly" Claudius Ptolemy, a genius for sure, developed orbits for the planets that included loop-de-loops or Ptolemy circles. Although the system he worked out was quite clever and creative, there is general agreement that planets orbit the sun, not the earth, and there are no loop-de-loops!

We have modern examples of believing something that is not true. Ulcers were thought to be caused by stress and treated accordingly as a chronic condition for decades. When it was discovered WHY most people get ulcers, that ulcers were caused by an infection with the bacterium Helicobacter pylori (H. pylori), suddenly a cure became possible.

Today we have some of our most brilliant people working on Ptolemy circles- creative answers for complex problems created by fallacies. The ancient world would never be able to reconcile cosmology with reality until they accepted the fact that the earth is not the center of the universe.

Is it possible that today there are monumental errors in our scientific thinking? Some of our great minds are proposing some loop-de-loops far more bizarre than Ptolemy circles.

For example, we have scientists saying that most of the material universe is missing. To make up for what is missing they invented dark energy and dark matter. Up to 95 percent of the universe they want us to believe is composed of this dark stuff. They did not discover dark matter and dark energy. It cannot be discovered because they say it is invisible, untouchable, and undetectable. Could it also be "make believe" like Ptolemy circles?

The dark matter theory is being put forward to explain gravitational effects that do not fit our current understanding of gravity and the Big Bang. Isn't it more likely that our understanding of gravity is wrong or the Big Bang is wrong? There is certainly room for error in our understanding of gravity since **we actually do not have an understanding of gravity.**

That is right. You may have been taught that Sir Isaac Newton explained gravity. You may know that Albert Einstein provided further understanding of gravity. Both men discussed what gravity does but not WHY gravity works. **"Why Theory"** offers a new look at gravity, an explanation of the mechanics involved, and even proposes a new name for gravity.

Another example of creating Ptolemy circles involves quantum mechanics. Trying to explain the irrational wave/particle duality phenomenon, great minds have come up with the "many-worlds" theory. Consider this direct quote from Wikipedia:

"The many-worlds interpretation is an interpretation of quantum mechanics... In lay terms, the hypothesis states there is a very large—perhaps infinite—number of universes, and everything that could possibly have happened in our past, but did not, has occurred in the past of some other universe or universes. The theory is also referred to as MWI, the relative state formulation, the Everett interpretation, the theory of the universal wave function, many-universes interpretation, or just many-worlds."

Really? Now I realize that you have to be pretty smart to create this "many worlds" theory- you have to be pretty smart to even understand what they are talking about- and even smarter to understand why scientists are desperately trying to replace the unbelievable "observer created reality" theory that Niels Bohr proposed.

But really, did someone seriously propose a "many-worlds" model and more surprisingly did anyone actually take this seriously? Just so you understand, this is like saying that when The New York Yankees win the World Series in our world, that same year every other American league and National league team wins the World Series in some 29 other worlds. Every team wins the World Series in the same year- including the Chicago Cubs. The Chicago Cubs, winning the World Series? Now you know there is something seriously wrong with that theory.

"Why Theory" will look at the wave/particle duality enigma and quantum mechanics from a different perspective. You may know that Albert Einstein disagreed with Niels Bohr over the interpretation and ramifications of the "observer created reality". The famous Schrodinger's cat thought experiment was on Einstein's side. Heisenberg's uncertainty principle supported Niels Bohr. After more than 80 years there is no explanation that "makes sense".

So let's begin our search for the missing WHYs in our scientific understanding. We will begin with magnetism.

MAGNETISM - WHY DOES IT WORK?

Do you know Richard Feynman? "Feynman became one of the best-known scientists in the world. In a 1999 poll of 130 leading physicists worldwide by the British Journal Physics World he was ranked as one of the ten greatest physicists of all time." -Wikipedia

There is a Richard Feynman video on YouTube appropriately titled "WHY". He is being questioned about magnetism. Dr. Feynman says, "I am not going to be able to tell you why magnets attract... I can't explain it in any terms. There is a magnetic force that makes them attract but I am not able to tell you why."

By the end of this chapter, you will be able to explain why magnetism works.

If you place a sheet of paper over a bar magnet and sprinkle iron filings on the paper, the filings will reveal lines of force coming out of one end of the magnet, bending back toward the other end of the magnet. (Diagram 1)

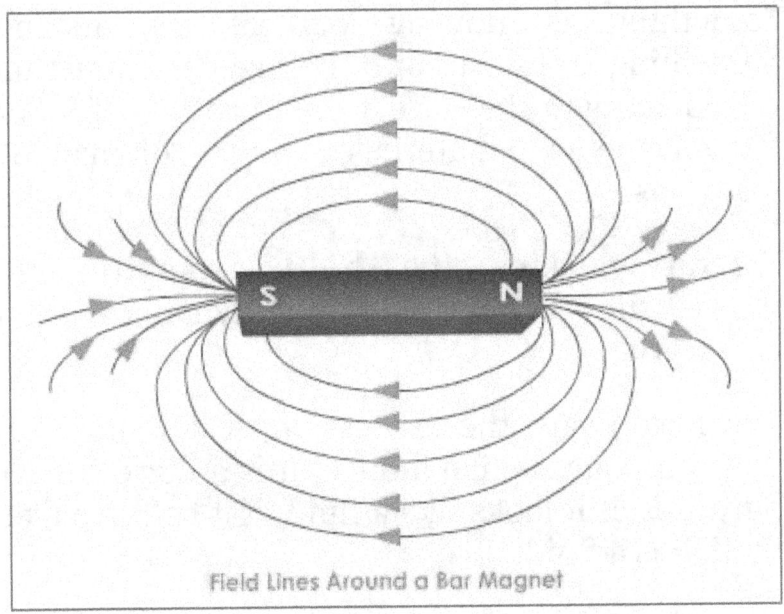

Field Lines Around a Bar Magnet

Like poles will repel each other- north to north or south to south.

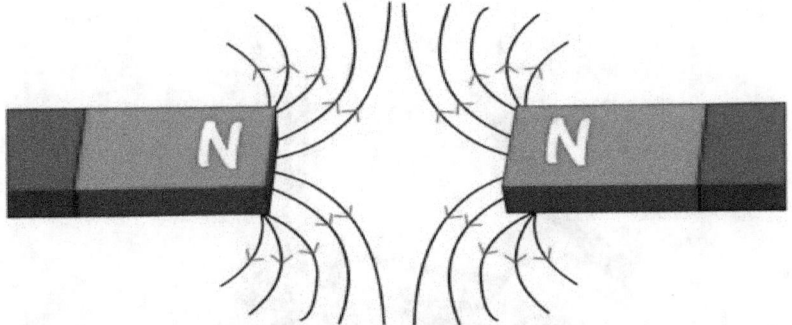

We know just about everything there is to know about magnetism except WHY any of this works.

Magnetism is not electricity- but it is a force. Those lines of force do not contain electrons but they do contain something.

Something is flowing out of the magnet. Something is flowing into the magnet. Something is affecting the electron orbits of both ferromagnetic materials and diamagnetic materials.

We are making the **bold assertion** that magnetism is explained by fluid dynamics, a fluid is at work.

You have heard the expression, if it walks like a duck and quacks like a duck, it is a duck. For our purposes, if it looks like a fluid, if it behaves like a fluid, it is a fluid.

To understand what is happening with magnetism we need to examine a common household appliance, a vacuum cleaner.

Shocker #1- There is no such thing as a vacuum cleaner. A vacuum cannot clean. **What we call a vacuum cleaner is a machine that creates a low pressure area inside the machine by using a fan to force air out one end.** Air is a fluid under pressure. At sea level the pressure measures 14.7 pounds per square inch.

Fluids flow from where there is more pressure to where there is less pressure. The air pressure outside the vacuum cleaner is greater than the air pressure inside the vacuum cleaner. Fluid dynamics explains why air rushes into the vacuum cleaner. The inflowing air picks up dust and dirt and pushes the dirt into the machine where it is filtered out of the air and trapped.

Shocker #2- There is no such thing as suction. I know you have used this word your whole life and no doubt will continue to use it.

Whether we are talking about a vacuum cleaner, a nursing baby, or you drinking a soda through a straw- there is no such thing as sucking or suction. Simply put, suction is supposed to be a pull, but that pull does not exist. Instead, there is a push. The push is from where there is more pressure to where there is less pressure.

Now back to magnetism. It is our theory that there is a non-molecular fluid that explains how magnets work. Since magnets work in outer space this fluid must be universal. The fluid is composed of spacetrons. You have never heard of spacetrons for a good reason. **Spacetrons, spacetron fluid, and spacetronics are newly minted terms that we will be using to describe WHY things work the way they do.**

Spacetrons cannot be measured or weighed. The existence of spacetrons is confirmed by the effects they have on matter, energy and magnetism. Spacetrons affect everything.

Normally spacetrons are scattered haphazardly by electrons like the tiny specs of dust in the air when you walk into a room. Is that a size comparison? I mean If we increase the size of a single electron to the size of your body, would a spacetron proportionately increase to the size of a speck of dust? The short answer is probably NO. The spacetron would still be unimaginably small, still impossible to measure. We will explain WHY in a later chapter.

In a magnet the atoms and their electrons are permanently aligned and organized creating a motorized fan effect similar to the fan in the vacuum cleaner. Instead of spacetrons being scattered in all directions inside the magnet, massive quantities of spacetrons are ejected from the magnet out of the north poll. This constant exiting stream of spacetrons reduces the "spacetron pressure" inside the magnet.

Suddenly, all the mysterious and magical actions of magnetism can be "scientifically" explained. For example, if there is a fluid coming out of the north poles of both magnets we can understand why the collision of those opposing flows would make the like poles repel each other.

We normally would say magnets attract iron, nickel, and cobalt. That is true but not for the reason we have been taught. We have been taught that a magnetic pull is at work. The opposite is true. The truth is these metals are pushed toward the magnet by spacetron pressure the same way dirt is pushed by air pressure into a vacuum cleaner. These ferromagnetic materials are strongly affected by spacetrons rushing toward the low spacetron pressure area created by the magnet.

Why are not all materials pushed by the spacetron stream toward the magnet? Non-magnetic materials are sometimes referred to as diamagnetic. Most materials are diamagnetic. These materials cannot be turned into a magnet like iron, cobalt and nickel so there must be something different about them.

Simply put, spinning electrons create a magnetic field with a positive charge and negative charge, like mini-magnets. In a magnetic substance all positive charges face the same way, all negative charges face the same way. In a non-magnetic substance, like copper and silver, they do not line up in the same orientation.

An incoming stream of spacetrons trying to pass through non-magnetic materials is scattered and the magnetic effect is lost.

WHY are there magnetic lines of force?

Lines of force are exactly what you would expect when you understand fluid dynamics. The emerging stream of spacetrons from the magnet meets resistance from the atmosphere of spacetrons. Fluids in this situation bunch up and start to turn. Since the reverse side of the magnet is a continuous low pressure area, the spacetrons are pushed in that direction.

Why lines of force?

Let me give you an example. Suppose seven or eight people are with you. Your group tries to pass through a mob of people. As your group pushes into the mob of people, invariably one person in your group will make some headway and the others in your group will automatically fall in line single file behind the leader. It becomes the path of least resistance.

As streams of spacetrons come out of the magnet they are blocked by a mob of spacetrons, an atmosphere of spacetrons, a spacetron fluid. Pushing through that resistance forces the spacetrons into similar lines creating your magnetic lines of force. The lines of flowing spacetrons creates the magnetic field.

Magnetic fields with their lines of force will appear in earth's atmosphere but they also appear in a vacuum and in outer space. WHY? After your best efforts to create a vacuum, it will still contain spacetrons. Spacetrons are impossible to remove. Spacetron fluid fills a vacuum. Spacetron fluid fills our solar system and our universe.

I realize we have made bold statements. The next chapter will explain the historical and scientific basis for accepting a universe filled with spacetrons.

MAGNETISM - WHY DOES IT WORK?

Significant revelations so far:

MAGNETISM is a PUSH not a PULL.

SPACETRON fluid pressure provides the PUSH.

SPACETRON FLUID, "THE ETHER" RESURRECTED, REVISISTED & REVISED

(Warning: There is a real shocker coming)

During the 1800s and early 1900s scientists were positive there was a fluid in the universe they called the ether.

In 1815, French physicist Augustin-Jean Fresnel (1788 to 1827) hypothesized that space is filled with a medium known as the ether because waves need something that can transmit them.

In 1869 Sir William Thomson who coined the phrase "kinetic energy" described the ether this way: "Space is continuously occupied by an incompressible frictionless fluid acted on by no force...." [TRSE, vol. 25, 1869, pg. 217-260, "On Vortex Motion"]

Waves of energy require something to carry the waves of energy. You cannot have waves in the ocean without the ocean. Sound waves travel in water and air. Scientists agreed that light waves from the sun needed something to transfer or carry these energy waves from the sun to the earth. Waves always require a medium.

However, in 1905 photons were discovered by Albert Einstein. Einstein explained the photoelectric effect by saying that "light itself is a particle," and for this he received the Nobel Prize in Physics.

This is where things got tricky.

Albert Einstein opined that the ether might not be necessary if light travels as photons. Think of photons as little bullets. A bullet from a gun, can travel through a vacuum. A bullet can travel through empty space.

When Einstein said the ether might not be necessary, Maxwell, Faraday, Kelvin, Fitzgerald, and Lorentz disagreed. These men were outstanding scientists of their day but their opinion was no match against the opinion of the world famous Albert Einstein. **This led to a generally accepted conclusion that the ether did not exist and that is the situation today, over 100 years later!**

There were famous dissenters like Charles Lane Poor, an American professor, who objected to discarding the ether. He stood in opposition to Einstein's theory of relativity as a whole.

Professor Poor wrote: "The supposed astronomical proofs of the theory [of relativity], as cited and claimed by Einstein, do not exist. He is a confusionist. The Einstein theory is a fallacy. The theory that ether does not exist, and that gravity is not a force but a property of space can only be described as a crazy vagary, a disgrace to our age".

But Albert Einstein was the "rock star" and these attacks were pretty much ignored. Even today, if you disagree with Einstein you run the risk of being considered a screwball, kook, crackpot, lunatic, weirdo... take your pick.

How dare you say the ether exists when Einstein says it doesn't? Who do you think you are?

(Here's the Shocker)

Einstein never said the ether does not exist. In 1920, lecturing at the University of Leiden, on the topic "Ether and the Theory of Relativity," these are his words:

"Recapitulating, we may say that according to the general theory of relativity space is endowed with physical qualities; in this sense, therefore, there exists an ether".

Albert Einstein continues:

"According to the general theory of relativity **space without ether is unthinkable**; for in such space there not only would be no propagation of light, but also no possibility of existence for standards of space and time (measuring-rods and clocks), nor therefore any space-time intervals in the physical sense".

Albert Einstein even wrote a letter to Hendrik Lorentz creator of the Lorentz ether theory (LET) making it clear that he was not contradicting his theory and acknowledging the existence of the ether.

This lecture was ignored. The letter was ignored.

The 20th century came to a close with the world "knowing" that the ether does not exist and that is the accepted view now in the 21st century.

"Space without ether is unthinkable". You have to admit that is a shocking statement coming from the man credited with removing the ether. We are hereby officially resurrecting the ether and giving it a new name "spacetron fluid". While the ether had a limited function, we intend to show how spacetron fluid plays a much bigger role in our understanding of the universe.

Spacetron fluid is composed of spacetrons. We are actually at a loss of words to describe spacetrons. A spacetron sounds like an electron, proton, or photon but it is fundamentally different. Those three are particles but a spacetron is not a particle.

Spacetrons are what particles are made of. More about this in the next chapter.

Before we leave the subject of the ether (spacetron fluid) we need to consider two other men, Georges Sagnac and Nikola Tesla.

Elon Musk named his luxury electric car company in honor of Nikola Tesla. Tesla's inventions include fluorescent lighting, the AC hydroelectric power system and wireless communication. Tesla was also a physicist who continued to believe in the existence of the ether long after Einstein's discovery of photons.

I was surprised while researching this book to discover a similarity between my theory of gravity and Tesla's ideas on gravity. I thought I was all alone with my unorthodox theory. It was good to find an ally with his credentials. Both of us explain gravity in a similar way, a way that will probably be completely new to you.

Do great minds think alike? The more you learn about Nikola Tesla the more you will agree that at least one of us is a genius.

Nicola Tesla is one of the 100 Scientists Who Changed the World

Georges Sagnac is a forgotten name in science. Wikipedia describes him this way: "Georges Sagnac (October 14, 1869 - February 26, 1928) was a French physicist who lent his name to the Sagnac effect, a phenomenon which is at the basis of interferometers and  ring laser gyroscopes developed since the 1970s".

After Edison's 1905 discovery of photons was being interpreted to mean the ether does not exist, <u>Sagnac designed a brilliant experiment that confirmed the existence of the ether.</u>

Nevertheless, the ether was banished from scientific discussions for 100 years. We have resurrected the ether, renamed it spacetron fluid, and are now prepared to use spacetron fluid to provide the WHY's that have eluded us.

Resurrecting the ether is tantamount to accepting the sun as the center of the solar system. Not recognizing the sun as the center of the solar

15

system was a scientific and religious blunder. Not recognizing the ether (spacetron fluid) has been an equally serious blunder.

The ether has been the missing piece of the puzzle. We have had to resort to fanciful stories and playing "make believe" in the world of science because the ether was discarded.

When science put the sun in its rightful place, the solar system and the universe started to make sense.

That is what is going to happen now because we have resurrected the ether. We are going to find the missing WHY's.

SPACETRON FLUID, "THE ETHER" RESURRECTED, REVISISTED & REVISED

Significant revelations so far:

> **Einstein did not deny the existence of the ether Einstein said the ether (spacetron fluid) is necessary for the propagation of light through space**
>
> **Georges Sagnac designed a brilliant experiment that confirmed the existence of the ether.**
>
> **Nikola Tesla used the ether to explain gravity in opposition to both Newton and Einstein.**

Albert Einstein: "Space without ether is unthinkable"

E = mc2 ......WHY?

Warning: There is another shocker coming

What do we learn from the world's most famous equation? We learn that there is a relationship between matter and energy. Albert Einstein's discovered that energy and matter are the same thing in different forms. Matter is concentrated energy. Energy is greatly expanded matter.

Does that description of E=mc2 sound about right to you? Well it is not. It contains a huge mistake, a blunder which we will explain a little later. First let's talk about matter.

Matter is composed of protons, electrons, and neutrons. Combinations of these three subatomic particles make up all the elements. The difference between one element and another is the number of protons in the nucleus of the atom. 47 protons- that's silver. 79 protons- that's gold. 82 protons- that's lead. Silver, gold and lead each has more neutrons than protons in the nucleus. Normally these elements have the same number of electrons as protons. Elements join together to form millions of different compounds. Everything you see, everything you own, everything you are, comes down to protons, electrons and neutrons.

But how does energy transform itself into electrons, protons and neutrons? No one has been able to answer this question...till now.

Now we have spacetron fluid to help us. We have resurrected the ether and renamed it spacetron fluid. Spacetrons are individual units of spacetron fluid (ether). Remember, 15 years after releasing his theory of special relativity, Albert Einstein said "Space without ether is unthinkable".

Think of the ether or spacetron fluid the way you think of earth's atmosphere. Earth's atmosphere is a mixture of multiple gases, oxygen, carbon dioxide, nitrogen, water vapor, and trace gases. The atmosphere is all around us. Earth's atmosphere is a molecular fluid and is confined to about 300 miles up around the earth. The weight of that 300 miles of atmospheric gases creates air pressure.

The ether or spacetron fluid is also all around us like an atmosphere. It extends into space and throughout space. It is not molecular which makes it virtually undetectable.

We are going to learn a lot about the spacetron fluid atmosphere by studying earth's atmosphere. Both are fluids- one molecular and one not. Both are governed by the laws of fluid dynamics.

In earth's atmosphere high pressure areas and low pressure areas develop. The rotation of the earth guarantees a constantly changing atmosphere. Temperature differences automatically occur between night and day. The air is always moving ranging from soft breezes to powerful winds.

Occasionally storms develop when winds begin to spiral. Hurricanes and tornados become visible, tangible and powerful. These storms are low pressure areas. The energy they possess comes from outside the storm. Fluids flow from where there is more pressure to where there is less pressure.

Storms concentrate enormous amounts of energy. Tornados can destroy buildings, homes, automobiles. The aftermath looks like a bomb has hit the neighborhood.

Hurricanes are much larger and more powerful than tornados. NASA says that "during its life cycle a hurricane can expend as much energy as 10,000 nuclear bombs!"

As hard as it may be to believe, all this energy comes from air pressure. Sunlight plays a part by making parts of the atmosphere warmer than other parts. This contributes to the creation of wind. When air moves its air pressure is reduced. This contributes to more air movement.

When storms hit land they unravel, lose shape, stop spiraling and release energy.

But why are we talking about the weather? Because there are parallels in the spacetron fluid atmosphere. In spacetron fluid there are in effect breezes, winds, storms, tornados and hurricanes.

Nuclear fusion creates the sun's energy which radiates continuously from the sun. Radiant energy takes different forms and has differing wave lengths and frequencies. We will discuss WHY wave lengths and frequencies occur in another chapter. The radiant energy we are most familiar with is light. Waves in the ocean are carried by the ocean. Similarly, light waves are carried by spacetron fluid. Since spacetron fluid provides very little resistance these light waves can travel over vast distances called light years.

We can imagine light waves moving in a linear fashion like the wind. Under certain circumstances energized spacetrons can be forced to spiral creating a spacetron storm. This most assuredly occurs in stars. Wherever a spiral occurs it has the potential of creating a low spacetron pressure area like a tornado or hurricane.

Photons, electrons, and protons are storms in the spacetron fluid atmosphere. These particles spin like tornados and hurricanes. One big difference: Protons, electrons, neutrons are permanent, almost indestructible.

(Here's is the Shocker)

E=mc2 does not mean what you think

E=mc2 does not mean energy becomes matter. It means that there is energy stored in matter. We are very familiar with calories (heat energy) stored in food. Calories refer to only electron energy whereas Einstein's equation refers to all the energy contained in electrons, protons, and neutrons.

Matter is NOT energy, it contains energy.

The formula is just a way of calculating how much energy is in a quantity of matter.

Subatomic particles are not made of energy. Energy does not become subatomic particles. Energy is necessary to create subatomic particles because energy is necessary to force spacetrons to spiral and compress. Once spiraling, these subatomic particles remain low pressure storms fed continuously by spacetron pressure. It is true if you remove all energy from matter, the matter will disappear. That is no different from what happens to a hurricane when the energy decreases and the storm stops rotating.

A hurricane contains enormous amounts of energy but the hurricane is not made of energy. **It is made of air and water.**

A storm's energy comes from air pressure outside the spinning storm. The spinning hurricane creates powerful winds up to 200 miles per hour but the spiral itself gets its strength from the incoming air because the hurricane is a low pressure area. As long as the hurricane is allowed to spin freely, it will remain a low pressure area and it will continue to be fed energy from outside air pressure.

Imagine a bizarre situation. Scientists decide that AIR does not exist. Now a meteorologist tries to teach how energy become storms, tornados or hurricanes, without mentioning air.

Silly, right? Something is missing. AIR. Obviously, you cannot have wind without AIR. Obviously you cannot have a hurricane without AIR.

We have a bizarre situation. Scientists mistakenly said there is no ETHER (spacetron fluid). Without spacetron fluid there is no way of explaining the existence or electrons, protons, and neutrons.

The energy in subatomic particles comes from spacetron pressure just like the energy in storms comes from air pressure. Spacetron fluid is to energy as air is to wind. Spacetron fluid is to subatomic particles as air is to hurricanes.

E = mc2 ......WHY?

Significant revelations so far:

The equation E=mc2 explains the concentration of energy in matter but not the existence of matter nor the existence of energy.

Energy is needed to make spacetrons spiral and keep spiraling.

Spacetrons are units of the spacetron fluid (ether).

Spiraling spacetrons become photons, electrons, protons.

The energy in subatomic particles comes from spacetron pressure.

GRAVITY... WHY IS THERE GRAVITY?

Warning: There is another shocker coming

A theory of everything would provide a unified explanation of the four fundamental forces- the Strong force - the Weak force - The Electromagnetic Force - and GRAVITY.

A theory of everything has been elusive. Not for want of trying.

We marvel at accomplishments of great scientists like Albert Einstein, Isaac Newton, Galileo, Nikola Tesla, Marie Curie, Louie Pasteur, Thomas Edison, Michael Faraday, James Maxwell, and of course many, many others. With primitive equipment, with limited access to information from around the world, with no computers, what these scientists accomplished is astounding.

These scientific giants have brought us very close to a complete understanding of our universe. You would think with our modern supercomputers, microscopes, telescopes, the internet, modern communications, the Cern supercollider, tens of thousands of scientists and billions of dollars for research, a theory of everything should be a piece of cake.

So far, we are stymied.

We have stubbed our TOE (pun intended). Something is seriously wrong. Something just does not make sense.

There are two reasons no theory of everything has emerged.

Obstacle 1. Science has accepted premises that are inaccurate. Trying to build an accurate theory on false premises is impossible.

Obstacle 2. GRAVITY.

No matter how brilliant Sir Isaac Newton (1642-1726) was in describing gravity and producing formulas for its calculation, he did not explain WHY gravity works. Three hundred years later we still do not have an answer to the fundamental question- WHY does gravity work the way it does?

It is time I introduce a Dutch mathematician who will help us finally explain gravity. Daniel Bernoulli was born February 8, 1700 and died 82 years later. It may seem strange that we have to go back more than 250 years to ask for Mr. Bernoulli's help.

You may have heard of the Bernoulli principle. He observed that boats were being pushed into a moving current of water by stagnant water. He thought that the fast moving water should have pushed the boats into the stagnant water. While trying to make sense of what he observed he discovered a universal law of fluid dynamics. Fluid pressure decreases as fluid speed increases.

An airplane wing works because of the Bernoulli principle. Air flows faster over the wing than beneath the wing. Air pressure under the wing is stronger than air pressure above the wing. That's it. That is the LIFT that enables a 747 plane to take off and remain in flight.

Daniel Bernoulli was 26 years old when the great Sir Isaac Newton died. The Bernoulli principle tells us WHY gravity works.

Spacetron fluid (ether) has a pressure. Spinning elemental particles composed of spacetrons are low pressure areas. Every spinning particle will require incoming spacetrons the same way a hurricane requires incoming air as long as it is spiraling. The earth is constantly absorbing a tremendous amount of spacetron fluid. The sun is absorbing massively more spacetron fluid.

Let's make this very simple. When Sir Isaac Newton saw an apple fall to the ground (no it did not hit him on head), he made the mistake of thinking the earth's gravity pulled the apple down. In truth, the apple was pushed down by the spacetrons streaming toward the earth. This influx of spacetrons is constant and proportionate to the quantity of atoms involved. This easily explains why the force of gravity is stronger on the earth than the moon and why the sun's gravity dwarfs them both.

(Here's is the Shocker)

Gravity is a PUSH not a PULL. Like suction, gravitational pull does not exist. Gravity, if defined as a pull, does not exist. To avoid future confusion, I propose replacing the term gravity with **spacetron influx.** Maybe after a while we can just say **influx.**

The sun's **influx** is greater than the earth's **influx**. All molecular bodies have **influx**.

Do you think this will catch on? It is unlikely I know. We still say "suction pulls" when we know "air pressure pushes". We still say sunrise and sunset when by now it should have sunk in that the sun does not rise or set- the earth rotates.

Bad science lingers. Look how the BIG BANG theory lingers. "(We will discuss that in another chapter).

We do not have to change any of Newton's formulas. We just need to change our thinking about them. For example, Newton said there was a mutual attraction between earth and sun. He said gravitational pull was the reason the earth orbited the sun. How do we explain orbiting planets now without gravity?

How does our understanding of spacetron influx explain what is happening?

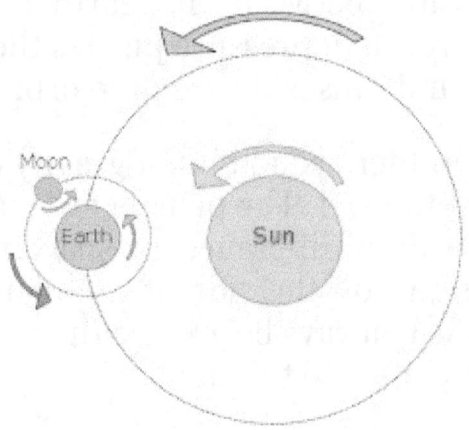

First it should be remembered that gravity is subtle. It is the weakest of the four fundamental forces. The WEAK force is actually stronger than gravity though it only operates for very short distances.

Massive amounts of spacetrons are being absorbed by both the sun and the earth. There is spacetron fluid between the earth and the sun. The pressure of that fluid is being reduced by a flow of spacetrons in two directions at once, toward the earth and a much greater flow toward the sun.

This double flow reduces the spacetron pressure on the side of the earth facing the sun. The spacetron pressure on the side away from the sun will therefore be stronger than the spacetron pressure between the earth and the sun. That difference in pressure pushes the earth toward the sun and causes the earth to orbit the sun.

Again there is no pull of gravity causing planets to orbit the sun. The planets are pushed into orbit by the subtle difference in spacetron pressure. The pressure on the side of the planet away from the sun will always be greater than the pressure on the side closest to the sun.

Replacing the pull of gravity with the push of spacetron influx should be easy for us to accept. Suction does not exist. Magnetism is a push, not a pull. Actually pulls are very hard to find. A friend of mine pointed out that you do not even PULL the trigger on a gun. Some say you squeeze the trigger. The truth is your finger is on the other side of the trigger and actually PUSHES the trigger mechanism back toward you.

What about Einstein's explanation of gravity as warped space-time instead of a gravitational force?

Both Einstein's interpretation of gravity and Newton's gravity predicted gravitational lensing. They both said that light passing by massive star would bend around the star. This would allow an

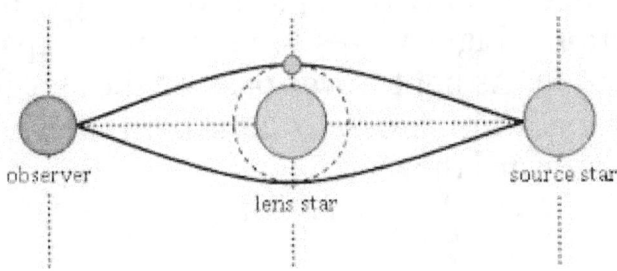

observer to see an object hidden by the star.

Einstein's general theory of relativity predicted the light would bend twice as much as Newton's theory of gravity predicted. During a 1919 total solar eclipse, Sir Arthur Eddington an English astronomer and his group performed the first experimental test of Albert Einstein's general theory of relativity. When Eddington announced Einstein the winner, general relativity became famous and Einstein became the "rock star".

But whether we are talking about Newton's gravity or Einstein's gravity what makes the light actually bend? WHY does light bend?

"Why Theory" states that stars are continuously receiving a massive spacetron influx from all directions. Light waves are streams of spacetrons in motion. As light waves pass by the star it is as if a strong wind crosses their path. Result- light bends toward the star.

GRAVITY ......WHY IS THERE GRAVITY?

Significant revelations so far:

Gravity is a PUSH not a PULL.

Spacetron influx or just INFLUX should replace gravity.

None of Isaac Newton's equations need to be changed

Bernoulli's principle explains WHY spacetron fluid flows continuously into planets, stars, and all molecular bodies.

WHY THE BIG BANG IS A BIG BUST

Warning: Shocker coming "the nail in the coffin"

I should not have to write this chapter. I am only discussing the Big Bang theory because **"Why Theory"** removes the basis for this theory. You should already know that the Big Bang Theory is a successful television show but a completely failed theory. It really is a mystery how this absurd theory ever gained any recognition. Absurd theory? Isn't that a little harsh and disrespectful?

Respect theoretical scientists but do not worship them. It is so easy to be impressed by someone presenting complicated formulas and scientific terms you do not fully understand. No matter how convincing that scientist may be, there are probably dozens of equally impressive intellects who disagree with him. Remember that a theory is a theory. There is a reason they call it a theory. Especially in cosmology, a theory is a supposition, and a supposition is an <u>uncertain belief</u>.

The 1992 book "The Big Bang Never Happened" by Eric Lerner exposed the weaknesses of the theory. The title says it all.

35

Ten years later the situation actually got worse for the theory. Meta Research Bulletin 11, 6-13 (2002) revealed "The Top 30 Problems with the Big Bang". Notice it says the "top 30". Makes you wonder if there was also a "bottom 30". A theory with two or three serious problems is usually discarded but The Big Bang theory keeps stumbling along. (With a popular TV show in international reruns, it may never die).

(Author's note: I recommend the following scholarly report on the defects in the Big Bang theory from the Pantheory Research Organization. I have no comment on The Pan Theory as an alternative to Big Bang theory).

June 30, 2015 the Pantheory Research Organization released an exhaustive report, **Problems with Big Bang Cosmology**, which focused on the top eight most serious problems:

- The Horizon Problem,

- The Anachronistic Galaxy Problem

- The Flatness Problem

- The Anachronistic Black Hole Problem

- The Density Problem

- The Metallicity Problem

- The Gravity Problem

- The Distance-Brightness Problem

The study also discusses problems with the "Ptolemy circles" being used to try to save the theory including dark matter, dark energy and the Inflation hypothesis.

Conclusion of the Study- Problems with Big Bang Cosmology

The summary conclusion of the paper is that the Big Bang model is probably the wrong model of cosmology, for reasons concerning the above and related problems, as explained within the paper.

The Inflation hypothesis is the most blatant example of just making it up as you go along. The Big Bang hit a dead end because the scientists calculated that even at the speed of light, the matter could not get far enough away to escape the gravity that would pull all the matter back. Result, sorry, no universe.

This should have been lights out for the Big Bang theory. Instead Inflation theory says that space itself was like a crumbled piece of paper that could stretch out faster than the speed of light. Oh really, and have you ever witnessed this phenomenon? No. Has this theory been tested? No. Can it be tested? No.

Then why has it been accepted? Because we need it to save the Big Bang theory. It is almost like saying let's make believe empty space did not exist before the Big Bang. Let's make believe that when the billions of galaxies containing billions of stars suddenly came bursting out of a "singularity", space itself was inside the same infinitely small and infinitely dense dot.

Didn't empty space exist before the singularity exploded? No, no, no, you are not even allowed to ask that question according to the theory's advocates. There is no "before" and no "space", because both time and space did not exist before the Big Bang. (This is when we finally say balderdash!)

What I find particularly appalling is the fact that scientists freely talk about the Big Bang, not as a discredited theory, but as an established scientific fact.

The CBS 60 minutes' program introduced a segment:

A Trip Inside The "Big Bang Machine" ... they reproduce conditions that hadn't existed since a tiny fraction of a second after the Big Bang." ... November 2, 2015.

Alternatively, there is the BBC documentary titled:

Big Bang Machine CERN Large Hadron Collider (LHC)

The machine and supporting budgets cost over $13.25 billion dollars. Scientist Austin Ball, who helped build it, said, ""Forty million times a second, bunches of protons collide in the center of this barrel section and they reproduce conditions that hadn't existed since a tiny fraction of a second after the Big Bang."

How did such a flawed theory get started in the first place? Edwin Hubble in 1929 discovered the light coming from far away galaxies in all directions was red-shifted. This was interpreted to mean that all these galaxies were moving away from us at enormous speeds. The further away the galaxy, the more red-shifted the light.

White light is made up of different colors- red, orange, yellow, blue, indigo, and violet. These are actually different wave length and frequencies. Our eyes and brain interpret each wave length and frequency as a separate color. The blue end of the spectrum has shorter wave lengths associated with higher frequency and higher energy.

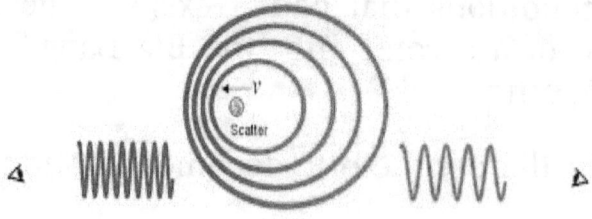

The redshift was explained using the Doppler Effect which says that sound waves and light waves will be distorted if the source of the waves is moving toward you or away from you. If a star or galaxy of stars is moving away from the earth at high speed, the light would be distorted producing a redshift. If the star or galaxy were coming toward you, the light would be blue-shifted.

Scientists are human. As humans they are not immune to the mind-numbing effect of group think and blind faith. Never has that been so clearly seen as in the adoption of the Big Bang theory. I cringe at article after article that casually mentions the Big Bang. Specifics, like the temperature of the universe within three seconds of the Big Bang, are embarrassing in their overreach and pompous certainty.

The Big Bang theory said that billions of years ago an explosion was responsible for an expanding universe. This must have been the greatest explosion in history- The Big Bang.

Cosmologists can mentally reverse the picture of an expanding universe, going back in time, contracting the massive universe, smaller and smaller until it all condenses into an infinitely dense "singularity". (There is a difference between something science has discovered, found, tested, measured, reproduced and something that is just a figment of someone's imagination. Until further notice when you see the word "singularity" you may substitute the word "figment").

Immediately, there is a huge problem with this picture and the interpretation of the Doppler effect. You can calculate the speed of the fleeing galaxy by how much the light is redshifted. Edwin Hubble found the further away the galaxy, the more redshifted the light.

But the Doppler redshift is not a measure of distance. It can be a measure of speed but then this picture means that for some bizarre reason the outlying galaxies are moving faster than galaxies closer to us. Billions of years after the Big Bang it appears that these galaxies are speeding up. **The "increased speed" of outer galaxies should be a warning sign that the Doppler effect is not causing the redshift.**

What else could be causing the redshift? How about light fatigue? Light is just getting tired after traveling distances that are so huge they have to be measured in light-years. Light fatigue has been considered, saying that collisions with space dust particles could be removing energy from the blue end of the spectrum and lengthening wave lengths (red-shifting).

Not bad but not perfect. If collisions with dust particles were causing the redshift, we would expect the images from these galaxies to be blurry but they are not. But wait, we are forgetting

41

something. The ether is back. Einstein said that the ether was necessary for the propagation of light. Spacetron fluid (the ether) is carrying light waves from these galaxies to our telescopic eyes.

Picture a rock falling into a pond. The waves it creates are strong and tight, the circles are close together. As the waves spread out, the circle gets bigger, the waves get flatter and the wave lengths get longer.

A wave in water is carried by water and is made of water. A light wave is made of spacetrons and is carried by spacetron fluid. Spacetron fluid is weightless and provides almost no resistance.

However, almost no resistance is still resistance and over vast distances the universal laws of fluid dynamics take effect. Light waves lose energy, wave lengths increase, and so light naturally redshifts.

Edwin Hubble really just discovered that light loses energy as it travels over vast distances. He did not discover an exploding universe.

In fact, in May, 2014 a team of astrophysicists led by Eric Lerner from Lawrenceville Plasma Physics reexamined the Tolman test for surface brightness dimming that was originally proposed as a test for the expansion of the universe.

Tolman test says in an expanding Universe the most distant galaxies should have hundreds of times dimmer surface brightness than similar nearby galaxies.

But that is not what observations show, as demonstrated by a study published in the International Journal of Modern Physics D.

The scientists carefully compared the size and brightness of about a thousand nearby and extremely distant galaxies. Contrary to the prediction of the Big Bang theory, they found that the surface brightness of the near and far galaxies are identical.

Eric Lerner's team declared that the Universe is not expanding at all. The Doppler effect therefore cannot explain the red-shift.

"the nail in the coffin"

The spacetron fluid model perfectly explains why the further away a galaxy is, the more redshifted the light. No need for accelerating galaxies billions of years after a Big Bang to explain the redshift. No need for the Big Bang at all.

When the Big Bang theory goes away, so do the Ptolemy circles of dark matter and dark energy. We do not need dark matter and dark energy to explain how the Big Bang was slowed down enough to allow for the formation of stars, planets, and galaxies.

We do not need dark energy to magically make far away galaxies accelerate billions of years after the Big Bang blast. The universe is not missing 95 percent of its matter. The universe is not expanding.

We do not need dark matter to put the brakes on the Big Bang runaway universe because as Eric Lerner was trying to tell the world 20 plus years ago, "The Big Bang Never Happened".

The Big Bang theory should have been dead and buried a long time ago. **"Why Theory"** just put **"the nail in the coffin"**.

WHY THE BIG BANG IS A BIG BUST

Significant revelations so far:

The Big Bang theory has been in serious trouble for decades.

The Doppler effect does NOT explain accelerating galaxies.

Light from distant galaxies is obeying laws of fluid dynamics.

Light naturally redshifts while traveling from distant galaxies.

The Big Bang never happened.

BEFORE YOU READ THIS NEXT CHAPTER READ THIS WARNING:

"One of the saddest lessons of history is this: If we've been bamboozled long enough, we tend to reject any evidence of the bamboozle.

We're no longer interested in finding out the truth. The bamboozle has captured us. It's simply too painful to acknowledge, even to ourselves, that we've been taken. Once you give a charlatan power over you, you almost never get it back."

— **Carl Sagan, The Demon-Haunted World: Science as a Candle in the Dark**

WHY THERE IS NO WAVE PARTICLE DUALITY AND NO OBSERVER CREATED REALITY

I assume if you have an interest in this subject, you have heard about the single slit / double slit experiment. Electrons sent against a board with a single slit act like particles.

We are approaching the century mark- 100 years- with no explanation of the wave particle duality mystery. A simple experiment had a surprising result. Surprise changed to shock when the experiment was fully understood.

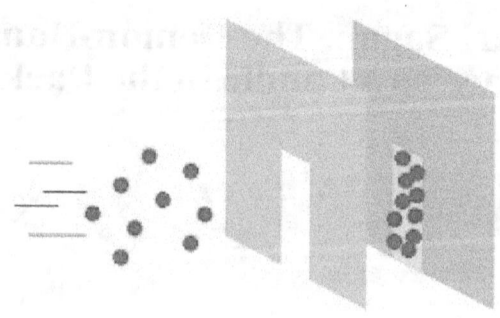

Electrons sent against a board with two slits suddenly act like waves.

The same thing happens with photons and protons.

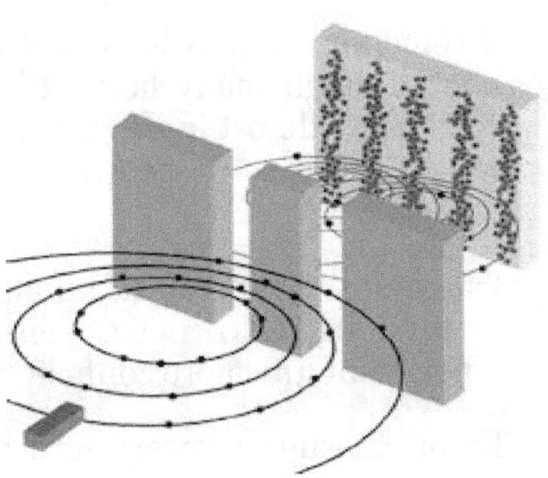

This is the wave/particle duality. So what's the problem? A particle is like a bullet and a wave is a transfer of energy. Why does putting an extra slit instantly transform a particle into a wave of energy?

After thousands of experiments and attempts to explain this transformation from particle to wave, scientists are still bewildered.

THEY SAY when electrons are being watched they are particles. When electrons are not being observed they are waves. The same is true for other subatomic particles. How can an observer change the world?

If you are confused, you are in good company. Werner Heisenberg who won the Nobel Prize for Physics in 1932 said this:

"I remember discussions with (Niels) Bohr which went through many hours till very late at night and ended almost in despair...Can nature possibly be so absurd as it seemed to us in these atomic experiments?"

The more the greatest minds of the twentieth century tried to make sense of this simple experiment, the more confusing things became.

Theories came forward utilizing the "observer created reality" premise. John Wheeler's Participatory

Universe theory boldly stated, *"There is no phenomena unless it is an observed phenomena."*

That is a profound statement and it means exactly what it says. There is no independent or absolute reality. Nothing happens until or unless someone or something observes it. Who is this John Wheeler? Sometimes you hear someone ask, "Is he anyone?" Well he is.

John Archibald Wheeler is a physicist. Wheeler worked directly with Niels Bohr, creator of the Copenhagen Interpretation. He pioneered the theory of nuclear fission. He worked on the top secret Manhattan project that produced the atomic bomb. "Black hole" and wormhole are terms he made popular.

For years Albert Einstein argued with Wheeler's friend, Niels Bohr, over whether there was an actual, real world or just assorted probabilities. But Einstein failed to defeat or replace the Copenhagen interpretation.

Finally, Einstein said:

"I still believe in the possibility of a model of reality—that is to say, of a theory which represents things themselves and not merely the probability of their occurrence."

Too bad Albert Einstein is no longer with us. I sincerely wish I could send him a copy of this book. We believe we are able to present the "model of reality" he wanted.

A model of reality requires two things:

1. **An explanation of the wave particle duality mystery.**

2. **Discrediting all of the "observer created reality" theories.**

"If quantum mechanics hasn't profoundly shocked you, you haven't understood it yet."

■ **Niels Bohr**

"The development of quantum mechanics early in the twentieth century obliged physicists to change radically the concepts they used to describe the world."

■ **Alain Aspect**

"Quantum physics is one of the hardest things to understand intuitively, because essentially the whole point is that our classical picture is wrong."

■ **Neil Turok**

The previous quotes give you an idea of the profound impact the wave-particle-uncertainty-principle-observer-created-reality has had on our thinking.

I once saw a magician take two spoons that fit together perfectly. He put one spoon down on the table and had an audience member gently hold the other spoon while the magician tried to bend the spoon mentally, without touching it- using just his mind. The magician 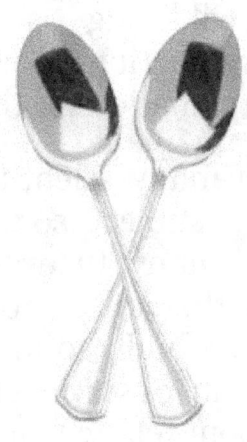 stared intently at the spoon being held about 12 inches from his face. We all focused on that spoon for about a minute.

The magician then asked the audience member to put the two spoons together again. They no longer fit together. Something strange had happened. We had a bent spoon. It was amazing. When you see something like that up close you are more than a little surprised. Wow. How did he do that? Did he really bend the spoon without touching it?

Normally I would never expose a magic trick but in this case the magician wanted us to know the secret. When he handed one spoon to the audience member, the magician simply bent the other spoon with his thumb as he was putting it down. When the two spoons did not fit, we automatically assumed the spoon we had been staring at was the one that bent.

I present this story to demonstrate how a very bewildering situation can have a simple explanation. Could there possibly be a simple explanation for the wave/particle mystery?

So many scientific articles have been written on this subject, so many books have been published, so many theories have been created, so many mathematical formulas, so many careers have been built, so much has been invested in the idea of an observer created reality that I feel terrible about telling you how the "magic trick" works.

The observer is not altering reality. How do I know that the observer is not creating reality? How can I be so sure? What earthshaking revelation do I have? What secret scientific principle have I discovered that has escaped the notice of thousands of scientists?

All those questions add up to one question, who do I think I am?

Let me answer that question by telling you about Bernard Sadow and Robert Platt. You probably never heard of them but they probably have helped you on every business trip or family vacation. These are the men who put wheels on luggage.

In 1970, Macy's sold for the first time, Mr. Sadow's four-wheel suit case that you could pull. In 1987 Mr. Platt invented two-wheel luggage with a long handle that could be pulled upright. Think of all the backaches these two men have helped mankind avoid. Next time you are at an airport, look around and see if you can find a suitcase without wheels. What a great idea!

The wheel was invented 5000 years ago by the Sumerians. Mankind has been transporting belongings in cases for at least 5000 years. How is it possible that it took till 1970 and 1987 before someone put wheels on luggage? What in the world took so long?

Were these two men smarter than the millions of men who lived before them? It struck me as really strange that automobiles, airplanes, rocket ships were all invented before wheeled luggage. Men walked on the moon before the first piece of wheeled luggage was sold!

How did all the major luggage companies and their professional designers and engineers miss the obvious benefit of combining wheels and suitcases? They must have asked themselves, "Why didn't we think of that?" The answer- they just didn't. Great ideas, thoughts, and discoveries often come from "unlettered and ordinary men".

Am I smarter than tens of thousands of scientists living and dead? The answer is NO but I know where my car keys are and they do not. It was just a child who noticed something was wrong with the king's wardrobe.

There is always a feeling of disappointment when you find out how a magic trick is done. Something that seems so impossible, so exciting, is usually explained by something so simple like "bending the other spoon". You miss the magic. You enjoyed being amazed. Right now there is magic involved in the wave/particle puzzle. There is even more magic involved in the "observer created" reality.

I do feel intimidated by the fact that the world's best and brightest scientists for almost a century have not been able to figure out "the trick".

I have something those scientists did not have. The ether. We resurrected the ether and renamed it spacetron fluid. As discussed earlier we believe subatomic particles like electrons are low-pressure storms spinning in the spacetron atmosphere.

I like the terms micro-world and macro-world because they are self-explanatory. The term quantum world has taken on a mystical aura along with quantum mechanics. We have come to believe that we have one set of rules and laws for the "real world" and another set unique to the mysterious quantum world.

Shocker: Particles do not become waves.

Particles do not become waves.

Everyone agrees when you set up the experiment with a single slit, photons, electrons, and protons produce a particle pattern. The particle pattern is like a fingerprint left only by particles. Proof positive.

Electrons always behave as particles when there is only one slit.

We have been taught that the double slit experiment changes particles to waves. Everyone "knows" that the observer creates the reality- when you have two slits, you see a wave pattern. Only waves can make a wave pattern. A wave pattern is a finger print, proof positive that the photons, electrons, protons are waves, not particles. Right? Wrong!

The current interpretation of the double slit experiment is that the observer creates the reality. A particle is a particle if you observe it as a particle. It is a wave if you observe it as a wave.

But something is terribly wrong with this interpretation.

The **shocker** is that electrons, photons, protons ALWAYS remain particles. If you have one slit or two slits or twenty slits- particles always remain particles. You may think I just made a mistake. Here is the proof I am not making a mistake.

I do not know why the following experiment by Hitachi is not more famous. It should be mandatory for every student. The scientists performed a double-slit experiment using electrons. You can watch the electrons coming in one at a time in this video produced by scientists at Hitachi in 1989. Printed below is a link to an internet video currently available. I hope it will still be available long after this book is published.

http://www.wired.com/2014/09/double-slit-empzeal/

The significance of the Hitachi experiment cannot be over-emphasized. They shot one electron at a time at a screen with the famous double-slit setup.

The experiment was supposed to prove something very spooky. It was supposed to prove that even shooting one electron at a time, you would still get an interference pattern. The conclusion was that each electron became a wave that could go through both slits. The electron wave coming out of the two slits could interfere with itself and produce a wave pattern. Let's put the wave pattern aside for the moment and look at this experiment.

One by one the electron hits the screen. The hit on the screen is proof positive that the electron is still a particle.

Amazingly the wave pattern does appear. BUT contrary to what everyone has been taught, contrary to what science teachers and university professors STILL teach, the wave/particle double slit experiment does not alter the nature of particles.

No one has been able to explain why electrons shot one by one create the wave pattern. But that is not the point. The point is the point. One by one, points appear on the screen, particle by particle.

Let that sink in for a while. We will explain the wave pattern but before we

do, please note- the observer is irrelevant. The observer is not changing a particle to a wave. There is no observer created reality. The entire observer created orthodoxy that has dominated for 100 years collapses like a house of cards.

The wave/particle double slit experiment does not alter the nature of particles. It only changes the distribution of particles. This is far less than what has been proposed, taught and accepted. But why is there a wave pattern?

When you walk into a room, you create air currents that you cannot see. When particles move through spacetron fluid, spacetron fluid currents are created. In addition, a stream of photons is flowing with a spacetron stream.

So it should not surprise us that a spacetron wave will accompany photons, electrons, protons, or even buckyballs (a molecule composed of a large number of carbon atoms).

When there is a single slit, there is no interference with the path of the particle even though spacetron fluid is accompanying the particle.

When there is a double slit, spacetron fluid flows through both slits, creates real waves that interfere with each other, and the accompanying photons, electrons, protons, small molecules, are "piloted" to the recording screen in a wave pattern.

BUT PARTICLES REMAIN PARTICLES AND WE HAVE THE PROOF IN THE NEXT DIAGRAM!

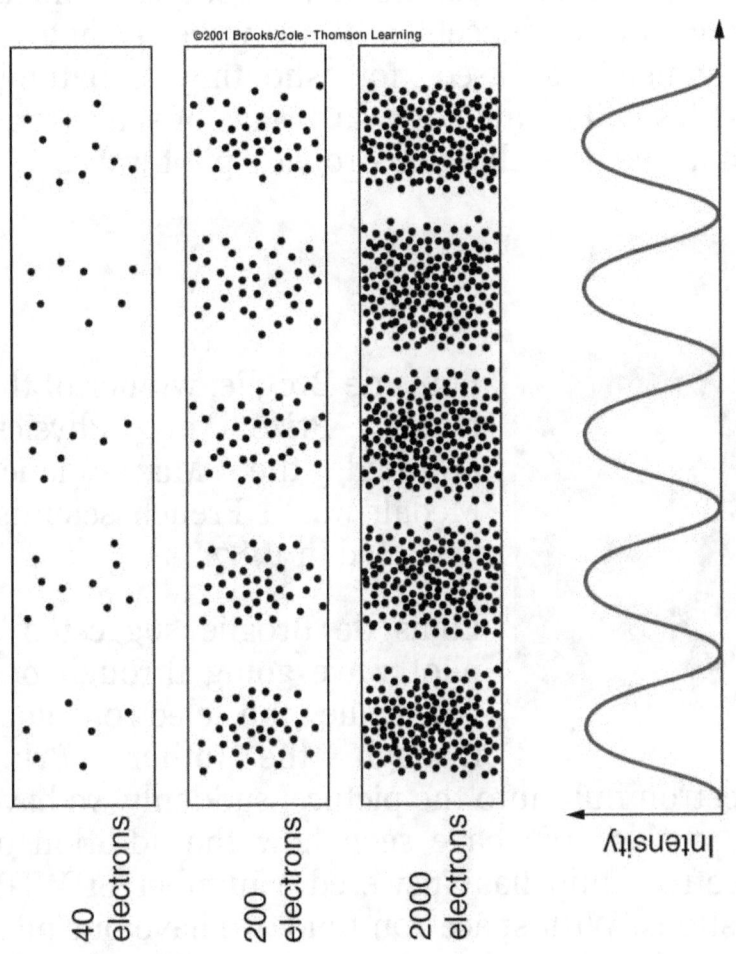

©2001 Brooks/Cole - Thomson Learning

40 electrons   200 electrons   2000 electrons   Intensity

We have already established that there is an atmosphere of spacetron fluid (the ether). It is all around us. Even in a vacuum chamber, spacetron fluid is present. When an electron is shot at our double-slit experiment it is like a shotgun blast. The photons, electrons or protons that are being propelled or shot at the double-slit experiment will create a spacetron fluid wave. Whatever mechanism is used for shooting subatomic particles will create a forward wave of spacetrons. We may be describing de Broglie's pilot wave.

Louis de Broglie, winner of the Nobel Prize in Physics, awarded the Max Planck Medal, was a French scientist who died in 1987.

Louis de Broglie suggested a pilot wave going through one slit while the electron goes through the other. Bring spacetron fluid into the picture, suddenly we have our answer. We have seen how the addition of spacetron fluid has answered a number of WHY questions. With spacetron fluid, we have our pilot wave going through both slits and influencing the distribution of photons, electrons or protons.

Now you know the magic trick. There is wave/particle partnership, NOT a wave/particle duality. The double-slit experiment results depend upon spacetron fluid influencing the distribution of particles. The fact that spacetrons waves can move particles and control where they land should not surprise us.

Spacetron fluid is at work in magnetism pushing objects around. Obviously spacetron fluid will have no trouble guiding photons or electrons. The most important thing to remember is the meaning of the Hitachi experiment- particles remain particles. There is no magical, mystical, observer caused transformation.

OBSERVER CREATED REALITY

We have learned so much from 19th century scientists. Now we can learn something from the 19th century humorist, Josh Billings:

"The trouble ain't what people don't know, it's what they know that ain't so."

Scientists who do not believe in ghosts and goblins became believers in a magical world where things go into and out of existence at will, where cats are alive and dead at the same time, where nothing really exists until it is observed. The observer creates reality simply by observing. What gave them this crazy idea? Let's see.

Brilliant scientists could not figure out why setting up an experiment one way produced a particle pattern while setting it up another way produced a wave pattern. In their minds, a wave cannot be a particle and a particle cannot be a wave.

Why does adding a slit in an experiment make this happen? They then developed theories and formulas that tried to describe what was happening and predicted probabilities.

None of the theories explained WHY this was happening. Just as a rule- never trust a theory that does not clearly tell you WHY something happens.

We now have explained WHY changing the format of an experiment produces the distribution of subatomic particles in a wave pattern. We have shown the most important truth, particles remain particles. There is no observer created reality. No magical transformation based on observation. Just fluid dynamics at work.

Particles are always particles. Spacetron fluid is always present to influence the double-slit experiment creating the pilot wave advocated by Louis de Broglie.

Interestingly, the double-slit experiment proves the existence of spacetron fluid.

BUT WE HAVE ONE MORE OBSERVER PROBLEM

The first illustration below shows what happens when subatomic particles are propelled against a double slit board with a recording screen behind. The recording screen shows multiple lines with spaces between the lines. That is an interference pattern. The interference pattern is proof positive that waves are at work but it does not prove subatomic particles are waves. Particles land discretely in a wave pattern because they are directed by spacetron fluid waves.

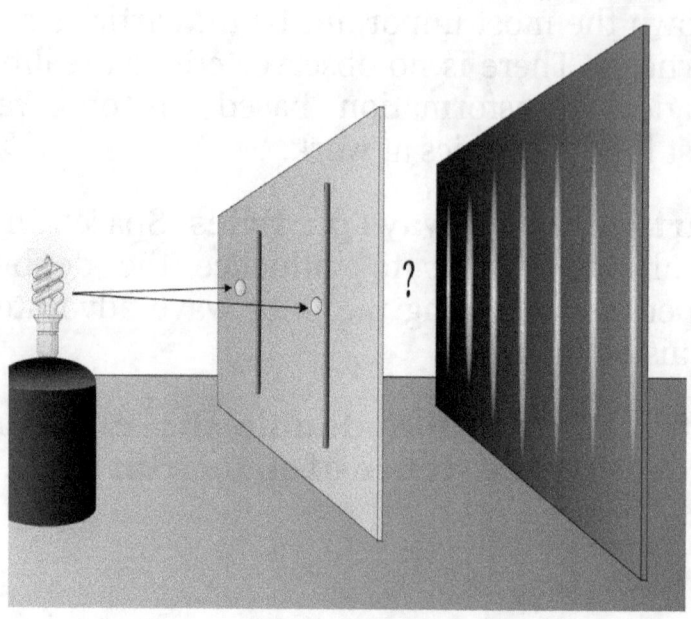

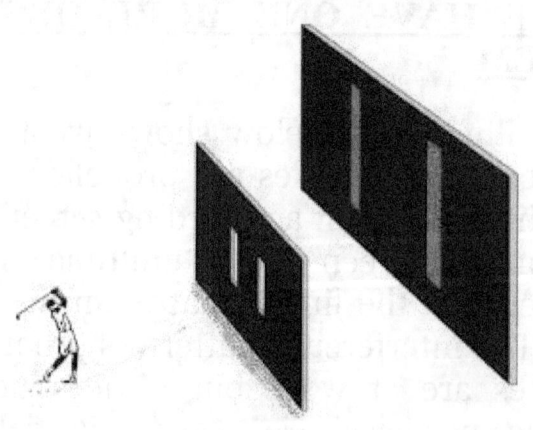

I have a very strange story to tell you. It involves the Observer Created Reality theory and Richard Feynman.

Richard Feynman was a very impressive physicist, writer and lecturer. He captured attention with his genius and personality. After Albert Einstein and Niels Bohr died, Mr. Feynman (1918-1988) became a prominent spokesman for quantum mechanics. He had a way with words and one of his quotes is particularly appropriate. *"The first principle is that you must not fool yourself and you are the easiest person to fool"*.

Richard Feynman proposed (as a thought experiment) that if detectors were placed before each slit, the interference pattern would disappear.

That statement "the interference pattern would disappear" has been accepted as fact, settled science. **But this was only a <u>thought experiment, not an actual experiment</u>.**

Before actual experiments could confirm or deny his expected result, educators and science writers added this "result" when teaching the double-slit experiment.

"Dr. Quantum" is a series of excellent teaching cartoon videos for children (and adults) on the subject of quantum mechanics. If you have had trouble grasping the theory and the issues involved, I highly recommend Dr. Quantum's videos. BUT BEWARE. After doing an excellent job demonstrating wave/particle duality, Dr. Quantum introduced the eyeball below. With the eyeball watching, the video shows the wave pattern disappears and is replaced with a particle pattern.

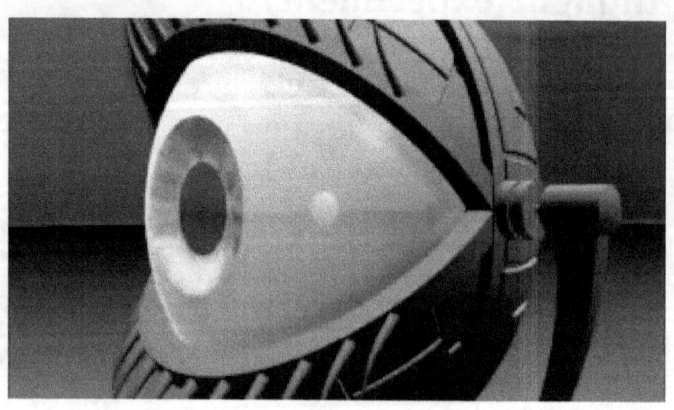

The eye symbolizes the idea that the quantum world is so weird that just looking at what is happening will change what is happening.

You are reading, **Why Theory**, so WHY did the brilliant Richard Feynman think that the wave pattern would disappear when you add a detector to the double-slit experiment?

If you are not already familiar with quantum theory, my answer is going to sound unbelievable. You see quantum theory believes the waves that cause the interference pattern are NOT REAL WAVES.

Waves in the ocean are real waves. Sound waves are real waves. The waves in the quantum world are probability waves. They think a photon wave or electron wave is really a "make believe wave", a probability wave, a theoretical wave.

They believe the waves are a picture of where an electron could be on its the way to the screen. It could be here so put a dot here or it could be there so put another dot there, then over here, etc. When you connect the dots, all the probable locations, you get a probability wave.

They believe that it is the convergence of two imaginary waves, mathematical probability waves, that produces the REAL interference pattern.

But Richard Feynman believed that just observing, just finding out where the electron actually was instead of where it could be would "collapse the probability wave". So the observer by observing would change the electron from a probability wave to an electron particle.

The Hitachi experiment accidentally proved that the pattern seen in the double-slit experiment is caused by particles. Single slit or double slit, particles remain particles.

Spacetron theory says that the wave pattern produced by the double-slit experiment is made by real waves, not probability waves. Spacetron theory says that ever present spacetron fluid is producing real waves and these real waves are forcing the electrons to be deposited in a wave pattern.

Real waves produce a real interference pattern. That pattern does not disappear because an observer is peeking. So we have two theories predicting radically different results. Which theory will be vindicated?

Shocker

"Observer created reality" has no observer

Earlier I challenged myself with these questions: How do I know that the observer is not creating reality? How can I be so sure? The answer is simple. In Mr. Feynman's thought experiment he did not say he was going to observe the electron passing through one of the slits because a human observer cannot see an electron or photon. So there is no observer. In Dr. Quantum's cartoon a robotic eye is shown but no such totally passive device exists- again no observer.

Sensors could be put in place but subatomic sensors do not work the way your eyes work. Most sensors use radiation- light, laser, infrared, radio waves, and ultrasonic waves. These sensors emit the radiation towards their target object and therefore can be destructive. Spacetron fluid is the lightest, flimsiest substance in the universe.

It is like the old joke, Question: How do you feed a rattlesnake? Answer: very, very carefully.

Richard Feynman said in his "thought experiment" a "Which-way" sensor that identifies which slit an electron or photon uses will magically make the interference pattern disappear and be replaced with the particle pattern. The premise is that the observer, even if it is just a mechanical sensor doing the observation, will change waves to particles. **The Hitachi experiment proves that particles do not change to waves. Therefore, waves do not change to particles.**

Fifty years after proposing his thought experiment it is still extremely difficult to develop a which-way sensor that would not interfere with the spacetron waves and wipe out the interference pattern. I believe a true observer, a non-destructive observer, would leave interference pattern intact while finding the path the electron took.

In fact, that result has been witnessed. Foundations of Physics reported in 1987 that information could be obtained regarding which path a particle had taken without destroying the interference pattern.

Spacetron theory says the interference pattern should remain in the double-slit experiment with or without sensors at the slits because spacetron fluid waves are real waves, not probability waves. The sensors must be delicately observing and not crudely disturbing the spacetron waves that are directing the photons or electrons.

If the spacetron waves are disturbed, the particle pattern will appear instead of the wave-interference pattern.

The ever present spacetron fluid has been interfering every time the double-slit experiment has been tried. Spacetron fluid acts like an unseen hand influencing the course a sub-atomic particle will take.

But it should be obvious that the sub-atomic particle is always accompanied by spacetron fluid whether there is one slit, two slits, or no slits.

The double-slit experiment exposes the existence of spacetron fluid. This is the same fluid that is responsible for those lines of force we discussed with magnetism. This is the same fluid that creates the "gravity effect" we call spacetron influx.

Why Theory

WHY THEORY - THE WAVE PARTICLE DUALITY AND OBSERVER CREATED REALITY

<u>Significant revelations so far:</u>

Subatomic particles are always particles.

Subatomic particles are accompanied by spacetron waves.

Spacetron waves are real waves not probability waves.

Single-slit experiment always results in a particle pattern.

Double-slit experiment always produces a wave pattern even though the particle remains a particle.

Spacetron fluid creates the interference pattern, the wave pattern in the double-slit experiment.

WHY DOES MATTER DESTROY ANTIMATTER?

To answer the question why does matter destroy antimatter we need to bring back a familiar picture. To be more accurate, the question should be why does matter and antimatter destroy each other.

Hurricanes in the northern hemisphere naturally turn counter-clockwise because of the coriolis effect. (If you fill a bathtub with water, when you release the water, the whirlpool will also turn counter-clockwise if you are in the northern hemisphere). Hurricanes in Australia you may have guessed turn clockwise. What do you suppose would happen if we could merge a southern hemisphere hurricane with a northern hemisphere hurricane?

We quoted NASA earlier saying that a good size hurricane releases more energy than thousands of atom bombs. So combining two hurricanes you might think that would produce a super storm, double the size.

That is not what would happen. You would end up with no storm at all. One plus one equals zero. The spin of one hurricane would conflict with the counter spin of the other. There would a tremendous release of energy but both storms would stop spinning. Hurricanes need to spin to exist.

It is not a coincidence that subatomic particles also spin- photons, electrons, protons. Antimatter consists of anti-protons, anti-electrons also called positrons. Their spins are opposite. Even neutrinos and antineutrinos, among the smallest known particles, have a half spin. They need to spin to exist. WHY?

Spacetron fluid (the ether) is an atmosphere. Energy moves spacetrons. Moving spacetrons carry energy. Light is a spacetron stream. Energy moves air- that is wind. Spiraling air becomes a tornado or hurricane. In the earth's environment, tornados and hurricanes have natural enemies that make sure they do not last long. Subatomic particles do not have natural enemies so they last indefinitely.

Spiraling spacetrons are subatomic particles. They exist in an atmosphere, spacetron fluid. Like earth's atmosphere, there is a spacetron pressure. Bernoulli's principle states that fluid motion reduces fluid pressure. Subatomic particles are storms that spin. In flowing spacetrons provide the energy for these particles. Just like a hurricane, to maintain low pressure, subatomic particles must spin. The spin creates a magnetic field, the electrical charge. The electrical charge is similar to the winds surrounding a hurricane.

Before **"Why Theory"**, was there any other explanation for subatomic particle spin? I do not believe so. Was there any explanation- WHY- subatomic particles have an electrical charge? I do not believe so. Now we have a WHY theory explaining spin and charge.

Quantum theory has an alternate description of spin. They say that spin does not mean spin. It is not really spinning. They have an alternate description of charge. But their theory fails to explain why or how it works. We have described what is happening using the laws of physics that govern the macro-world. They govern the micro world as well.

Before quantum theory decided that subatomic particles exist in a state of fuzzy probabilities, this was the classical view. A proton is 2000 times larger than an electron so obviously the diagram was not drawn to scale since electrons and protons are incorrectly shown as the same size.

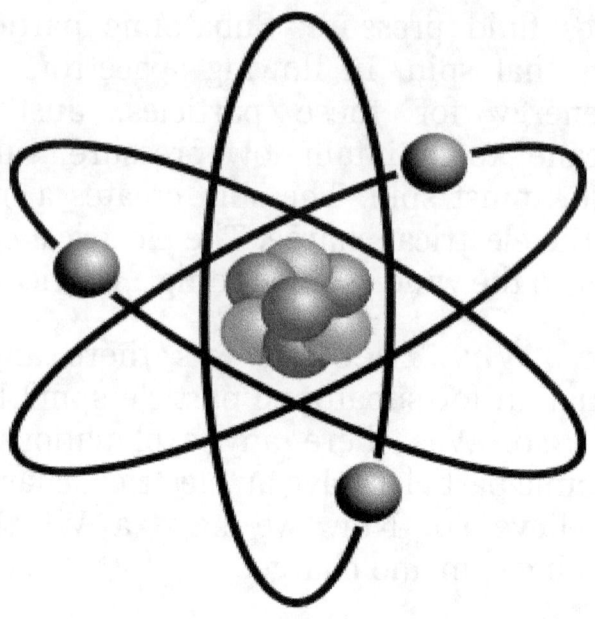

Now theory atom this

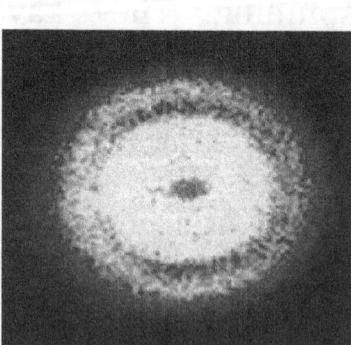

quantum depicts the way:

Twenty years ago when I suggested to the Russian scientists that we could use the energy from a matter / anti-matter reaction, I had no idea how silly my idea was.

Antimatter does not naturally occur in our world. It can be created at tremendous, mind-boggling cost. One estimate I saw said that it would cost $25 billion per gram or about $100 billion for a teaspoon of antimatter.

We really do not know why our universe does not contain 50 percent matter and 50 percent antimatter. This is just wild speculation but perhaps we are in the equivalent of the "northern hemisphere" universe. Maybe there is a "southern hemisphere" universe where there is only antimatter.

(I have always wondered why right-handedness and left-handedness were not 50/50 instead of 90/10).

It is a good thing we do not have antimatter around here. One gram of antimatter, 1/4 of a teaspoon, would release more energy than three Hiroshima strength atom bombs.

The question "why does matter destroy antimatter?" is not difficult to answer. A counter-clockwise northern hemisphere hurricane merging with a clockwise southern hemisphere would cause both to stop spinning. The release of energy would be enormous and the hurricanes would disappear. There are not a lot of ingredients in a hurricane, just air, water, and energy.

We have just described what would also happen if an antiproton and proton merged. The spin versus counter-spin would stop both from spinning. We can think of subatomic particles as tightly wound springs.

When a particle stops spinning all the energy used to wind up that spring is suddenly released. E=mc2 was a way of expressing the enormous amount of energy bound up in any particle of matter.

The release of energy involving any large amount of matter and antimatter would dwarf atomic bombs which only release a fraction of the energy in mass.

Scientists have changed the way they think about SPIN. Quantum theory needed to be adjusted after accepting the observer created reality. **"Why Theory"** says the classical view of spin and the structure of atoms with orbiting electrons needs to be restored.

"Why Theory" says that all subatomic particles are spinning storms in the spacetron fluid atmosphere. Adopting this structural view of atoms explains why matter and antimatter annihilate each other. We are not talking about a casual, lazy spin. These spins are at super speeds. Nicola Tesla calculated that energy speeds can actually exceed the speed of light although this is controversial.

Concrete proof that relativity can be violated can be found in George Gamow's watershed book Thirty Years That Shook Physics.

Gamow, one of the founding fathers of quantum physics, tells us that in the mid-1920's, Goudsmit and Uhlenbeck discovered not only that electrons were orthorotating, but also that they were spinning at 1.37 times the speed of light.

How important is spin to matter? Spin formed the particle in the first place. Spin maintains the low spacetron fluid pressure necessary for the particle to stay together. Spin produces the magnetic field.

We know that the negative electric charge of an electron equals the positive charge of the proton even though a proton is 2000 times more massive than the electron. We believe the proton spins at a much slower rate than the electron allowing the electrical charges to be equal.

Our illustration of two hurricanes merging with opposite spins gives a picture of what happens in the subatomic world. The laws of physics that rule the "real world" rule the quantum world as well. A proton and antiproton will stop spinning, releasing the energy that held the particle together. Neutrons match up with antineutrons. Electrons match up with anti-electrons. The match ups are mutually destructive.

Final result- the energy is dispersed and the spacetron fluid returns to the spacetron atmosphere.

WHY DOES MATTER DESTROY ANTIMATTER?

Significant revelations so far:

Classical view of the atom returns, with electrons orbiting a nucleus of protons and neutrons.

All particles are storms made of spiraling spacetrons.

Each particle has an antimatter counterpart. Particles, like storms, must keep spinning.

Spiraling subatomic particles contain enormous energy.

When each particle engages its antiparticle both stop spinning.

When particles stop spinning there is a rapid increase in spacetron pressure, an explosive release of energy, and a dispersion of spacetrons.

Matter and antimatter annihilate each other.

WHY IS THERE A STRONG FORCE?

(of course there is another shocker coming)

The picture shows a strong man, Sisyphus, pushing a huge stone up a hill. The energy it takes to move that stone is stored in the stone. When Sisyphus releases the stone, get out of the way.

The nucleus of the hydrogen atom has one proton. That is why hydrogen is the lightest element. Helium has two protons. It is the second lightest. Uranium has 92 protons in its nucleus. Uranium 238 also has 146 neutrons trapped in the nucleus of each uranium atom. Uranium with 3 less neutrons is the famous U235 which is unstable.

The force that holds the nucleus together is called the strong force and it has to be strong. Protons naturally repel each other. 92 protons do not want to be herded together. Neutrons are only stable in the nucleus. Outside the nucleus, the neutrons will decay, eventually to a proton, electron and an anti-neutrino. Holding all this together is the strong force. But where does the strong force come from?

We changed the name of gravity to spacetron influx and showed that the "force of gravity" is not a pull but rather a push caused by spacetron pressure. Spacetron pressure produces spacetron influx. Influx produces all the effects formerly attributed to gravity.

Is it possible that the strong force is also derived from spacetron pressure? WHY is there a strong force? Have you ever heard a WHY explanation for the strong force?

There seems to be nothing in the nucleus of the atom generating this strong force so the force must come from outside the nucleus.

We can learn from air pressure. We do not feel air pressure because our body pressure equals the pressure outside. You may be surprised to learn a basketball has over 4200 pounds of air pressure pressing in on its surface area. Your head would be crushed by thousands of pounds of pressure if the equal pressure inside your head disappeared.

The strong force is proportionate to the mass of the nucleus. The more protons and neutrons in the nucleus the greater the strong force. Smaller nuclei have less strong force than larger nuclei. WHY?

Every spinning neutron and proton is a powerful motor that reduces pressure inside the nucleus. Universal spacetron pressure becomes focused on the nucleus. Scientists have been trying to combine the gravity with the strong force- we have just done it. Actually we have combined the global influx that produces the gravity effects with the strong force.

(Here is the Shocker)

The strong force and "gravity" (influx) are one and the same.

You may be wondering if is this is an oversimplification. We have only been focusing on protons and neutrons. You may be wondering about Quarks. Although they have never been seen, there is evidence that both protons and neutrons have three quarks each. The evidence is far less credible than commonly believed. If quarks exist, then we are talking about wheels within wheels. Quarks spin. Everything that spins creates influx. **Strong force is concentrated spacetron influx.**

In order to explain how the strong force works, scientists theorized that there were particles bouncing between the quarks. Coincidentally, these theoretical, massless, particles were called GLUONS. The name sounded like glue which fit the idea that these particles were mediating the strong force, holding things together. **"Why Theory"** says that the strong force does not originate inside the atom. **The strong force is concentrated spacetron influx, an external force that holds the nucleus together. There is no need for GLUONS.**

Gluons may not exist.

THE WEAK FORCE / SHOCKER

We are ready to take a look at the **WEAK force**. You know that some materials are radioactive. When an element is radioactive it means that energy is leaking from the nucleus. Neutrons in the atom may be breaking up releasing a high energy electron from the nucleus. This is beta radiation. Alpha radiation occurs when protons and neutrons escape from the nucleus.

It is believed that the weak force causes radioactive decay of subatomic particles in the nucleus. The W boson and Z boson are two theoretical force carriers of this weak force.

A boson is a force carrier. Photons for example are force carriers. Photons are absorbed by electrons and photons are emitted from electrons. In the process energy is being carried by these photons. They are force carriers. Force carriers are bosons.

The W stands for the weak force and the W boson is supposed the be the force carrier that causes the nuclei of unstable elements to decay. (The Z boson is supposed to be just like a W boson but it has a neutral electrical charge while the W boson can be positive or negative). It is estimated that this boson is 100 times more massive than a proton and a proton is almost 2000 time more massive than an electron. That makes these bosons massive but there is a problem. These descriptions are theoretical. No one can hand you a jar of W or Z bosons. The discovery of the W and Z bosons themselves had to wait for the construction of a particle accelerator powerful enough to produce them.

There is controversy over the ever expanding list of particles "discovered" in high energy particle collision experiments.

"Why Theory" boldly asserts that the W boson and the Z boson do not exist in nature regardless of whether or not they can be produced by high energy collision.

When particles collide in accelerators, the energies released can give rise to hundreds of transient particles. These decay into more stable particles in a fraction of a second. There are no W bosons or Z bosons to take away from that process. There are no experiments that can be done with these particles. There are no practical applications for these particles. These particles have something in common with quarks and gluons- not a single one of these supposedly physical particles has ever been directly observed.

Trying to explain the weak force gets very complicated and involves the particle zoo, including six types of quarks, three types of bosons and three types of neutrinos.

THE STANDARD MODEL

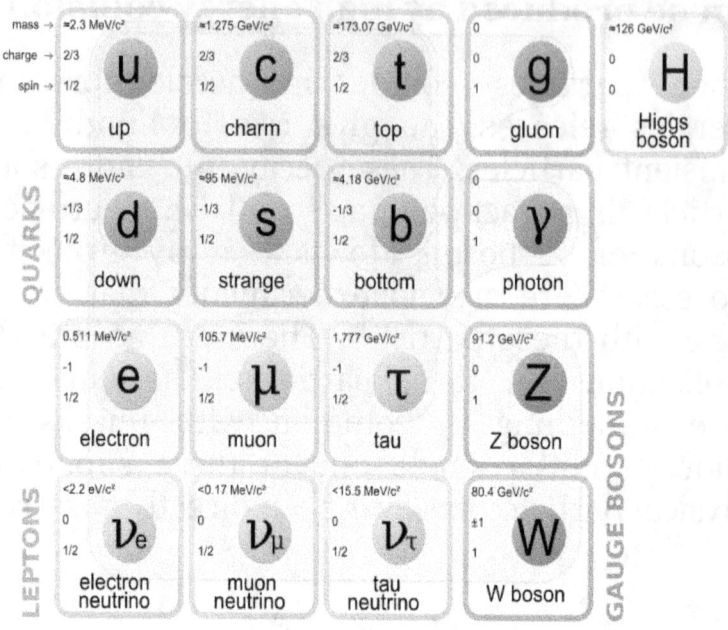

There is a much simpler explanation for radioactive decay or nuclear decay and radiation that does not involve the weak force and weak force carriers (negative and positive W bosons, and neutral Z bosons).

There are two types of radiation involved in nuclear decay- alpha radiation, beta radiation. When Uranium-238 decays into Thorium-234, an alpha particle is released from the uranium nucleus in the form of alpha radiation.

The alpha particle is a molecule of helium with two protons and two neutrons. Whenever a proton or protons leave the nucleus, the element changes to a completely different element.

When the uranium nucleus loses 2 protons it becomes thorium. It also loses 2 neutrons so uranium-238 becomes thorium-234. This is an example of alpha radiation which occurs when protons leave the nucleus of any element.

The number following an element is the total number of protons and neutrons in the nucleus. Carbon-14 for example has 6 protons and 8 neutrons. Carbon-14 decay is an example of beta radiation decay. The beta particle that escapes the carbon-14 nucleus is a high energy electron.

When that electron is ejected from the nucleus, the neutron it came from becomes a proton. The electron leaves the nucleus as radiation while the proton remains. Carbon-14 now has one more proton and one less neutron so it becomes Nitrogen-14 with 7 protons and 7 neutrons.

Now let's use the magic word WHY. Why is carbon-14 radioactive and carbon-12 is not? Carbon-12 has six protons and six neutrons in the nucleus and is stable. Why should the presence of two extra neutrons make carbon-14 radioactive? If the weak force is universally present like the strong force then all elements should be experiencing decay. All elements should be radioactive.

The nucleus is under pressure, being squeezed by the strong force. Protons repel protons. Neutrons outside the pressure of the nucleus automatically decay into protons and electrons. With the right combination of protons and neutrons, like the pieces of a puzzle fitting together, the nucleus is balanced and the strong force has no trouble locking everything in place.

Think of the nucleus as spring-loaded. When the mixture of protons and neutrons becomes unbalanced, the strong force may fail to hold it all together. The nucleus "springs a leak". It is as simple as that. No need for the weak force.

The tires on your car normally maintain air pressure while standing still or speeding at 80 miles per hour. Occasionally a weakness develops and a leak occurs. The pressure inside the tire will force air out.

Scientist have detected protons, neutrons, neutrinos, electrons leaking from nuclei of radioactive elements. They never detect W bosons or Z bosons. The strong force contributes to the expulsion of decaying particles from the nucleus. No weak force is necessary.

Considering the enormous spacetron pressure holding nuclei together, it should not surprise us that when a weakness develops particles come shooting out of the nucleus. With the wrong mix of neutrons and protons, the nucleus becomes unstable. Internal pressure caused by the strong force is at work. The natural tendency for neutrons to become unglued and the repulsion of positively charged protons add to that internal pressure.

There is no need for the Weak Force to explain atomic decay or radiation. Our theory of everything, our grand unified theory does not have to include the Weak Force because it does not exist. SHOCKER.

WHY IS THERE A STRONG FORCE?

<u>**Significant revelations so far:**</u>

The strong force was thought to reside in the nucleus.

Gluons are theoretical particles said to mediate the strong force.

The Strong force is intense spacetron influx.

The Strong force is an external force, not internal.

The strong force and "gravity" (influx) are one and the same.

Gluons are not necessary to create gravity. Gluons may not exist.

"Why Theory" says there is no Weak force.

Radioactive element decay is caused by failure of the strong force to hold unstable nuclei together.

SPACETRON THEORY: WHY ARE THERE LIGHT WAVES?

Why are there light waves? What a simple question. If you think you know the answer, try answering it. I mean, put your answer into words. Not so easy is it? You probably have never been asked that question before.

I will focus the question better. If photons are like bullets why don't they shoot straight from the sun to the earth? What is all this up and down stuff along the way? We are going to tackle the WHY question, "why are there light waves". It is much more challenging than you might think. We will be bumping up against the "conventional wisdom" again.

The current explanation is a bundle of confusion caused by misunderstanding the apparent wave/particle duality. We have shown that there is a wave/particle 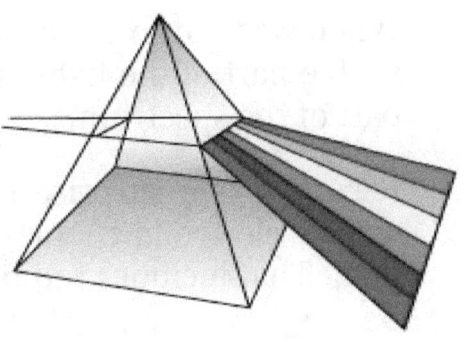 partnership, a wave and a particle.

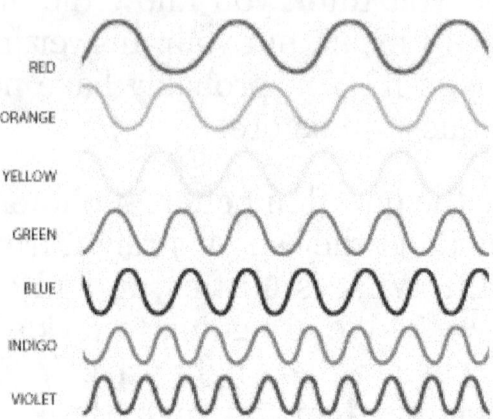

When we mention light waves we include both the visible part of the light spectrum and the invisible part of the spectrum.

It comes as a surprise to many to learn that the entire electromagnetic spectrum is considered light. That includes radio waves to gamma waves.

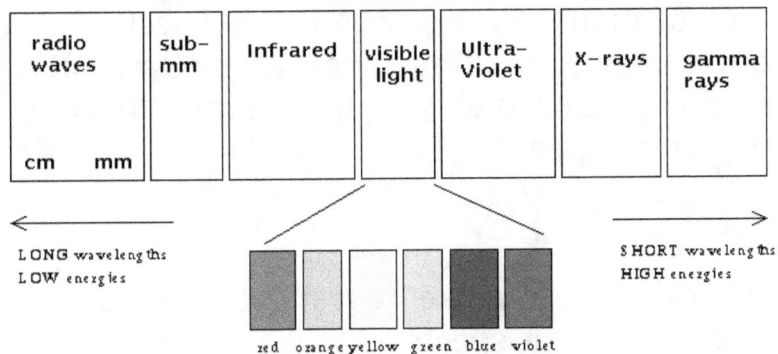

Radio waves have the longest wave length. The shortest wave lengths are gamma waves. The shorter the wave length the greater the energy. This fact is critical to understanding why there are light waves at all.

The electromagnetic spectrum consists of gamma-rays, x-rays, ultraviolet rays, visible light, infrared rays, microwaves and radio waves.

These seven categories of light are fundamentally the same. While most people think of photons exclusively with visible light the full spectrum exhibits the wave/particle *partnership AND THE PARTICLE IS THE PHOTON!*

The basic unit of energy is the photon. The basic unit of matter is the proton. The electron is the other major building block of matter. It is a marvel that on such simplicity our entire complex universe depends.

WHY does light make waves? Why does the entire electromagnetic spectrum make waves?

The sun radiates an enormous amount of photons in all direction and in all possible levels of energy.

The photons are not being dumped into empty space because - as we have been saying - there is **no such thing as empty space**. These photons are thrown into spacetron fluid. Even at the highest energy level photons will meet some resistance from the fluid.

The resistance is very slight but the lead photons will slide slightly off course while the trailing photons will push the leading photons. This fluid stream of photons is moving at the speed of light. How fast is the speed of light? In 1/7 of second, light could travel once around the world, 25,000 miles.

This is where fluid dynamics and spacetron pressure provides the magic needed to produce waves. Bernoulli's principle is at work again. As fluid moves faster the fluid pressure decreases.

In this case the photons and spacetrons stream moving at the speed of light has almost zero pressure. The surrounding spacetron fluid squeezes in on this photon stream. The effect is to produce a "push back" and wiggling effect.

These are real waves, not probability waves. The peaks and valleys are caused by the photons bouncing off the high pressure surrounding spacetron sleeve.

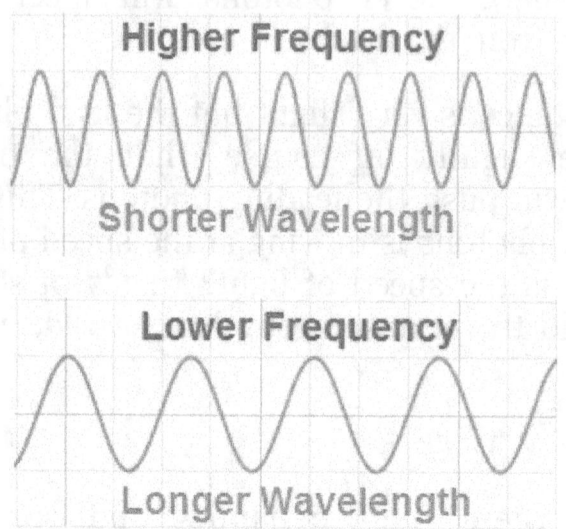

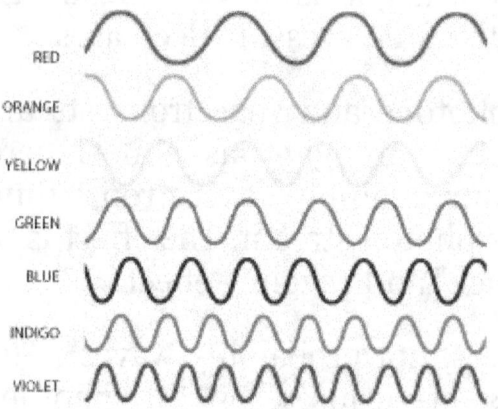

The colors have different wavelengths, frequencies, and energies. Violet has the shortest wavelength in the visible spectrum.

That means it has the highest frequency and energy. Red has the longest wavelength, and lowest frequency and energy.

As Albert Einstein said: "According to the general theory of relativity **space without ether is unthinkable;** for in such space there not only would be no propagation of light, but also no possibility of existence for standards of space and time (measuring-rods and clocks), nor therefore any space-time intervals in the physical sense".

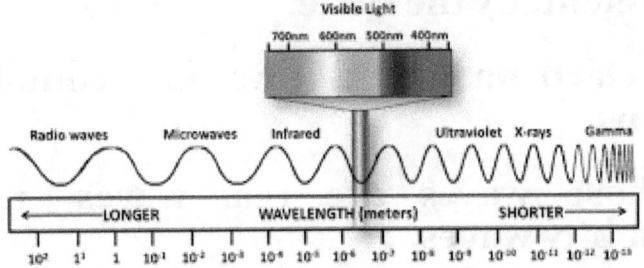

We now see how spacetron fluid (ether) is responsible for the propagation of light. From radio waves to gamma waves, spacetron fluid pressure is responsible for energy waves in space.

SPACETRON THEORY: WHY ARE THERE LIGHT WAVES?

Significant revelations so far:

There is a visible light spectrum.

There is also an invisible light spectrum

These seven categories of light are fundamentally the same.

All electromagnetic waves contain photons.

All those waves are real waves- not probability waves.

Streaming photons become trapped in a spacetron pressure sleeve.

The peaks and valleys are caused by photons bouncing off the high pressure surrounding spacetron sleeve.

FINAL THOUGHTS

Be Smarter than Isaac Newton, Niels Bohr,

Nicola Tesla, Albert Einstein,

Stephen Hawking and everyone you know

"Why Theory"

"One of the saddest lessons of history is this: If we've been bamboozled long enough, we tend to reject any evidence of the bamboozle.

We're no longer interested in finding out the truth. The bamboozle has captured us. It's simply too painful to acknowledge, even to ourselves, that we've been taken. Once you give a charlatan power over you, you almost never get it back."

— Carl Sagan, The Demon-Haunted World: Science as a Candle in the Dark

If you are familiar with the "conventional scientific wisdom" you realize that this book is not just a clarification or expansion of that wisdom. You may consider it an all-out attack. We are not rocking the boat. We are sinking it.

By simply demanding WHY ANSWERS, we were forced to confront the modern Ptolemy circles that have tried repeatedly to bolster failed theories.

We mentioned the five friends, What, Where, When, How, and Why. Any theory that does not tell you Why is incomplete, suspect and open to attack.

I discovered that WHY ANSWERS were missing for many of the most basic scientific questions. Then the key that opened one door, turned out to be a master key for many doors.

Einstein said, "I still believe in the possibility of a model of reality—that is to say, of a theory which represents things themselves and not merely the probability of their occurrence."

I did not set out to produce the "model of reality" that Einstein so desperately wanted. I used the word SHOCKER several times in this book to call the reader's attention to some new or unconventional idea. The biggest SHOCKER came when I realized that **"Why Theory"** is that "model of reality" that Albert Einstein so desperately wanted.

As Carl Sagan's comment says, it may be impossible to get the scientific community to even consider ideas that are so destructive to conventional wisdom. Copernicus and Louis Pasteur met violent resistance to their "way out" theories. Galileo was tried by the Inquisition, found "vehemently suspect of heresy". He was forced to recant.

What does that mean "forced to recant"? It means he had to say he no longer believed the ideas the church found to be heretical or he would be burned to death. Even though he did recant he still spent the rest of his life under house arrest.

I am pretty sure I will be able to avoid the drastic response Galileo received. I am more concerned that some may find excuses to ignore the basic truths we have discovered and discussed. This book was written for everyone, not just scientists and college professors.

It is my hope that **"Why Theory"** will inspire further discussion and ignite real interest in this area of science.

We look forward to receiving your thoughts. If you send me your email address I will share with you information on the latest discoveries and feedback.

Just send an email to science@inncareers.com.

At a time when the applied sciences are producing mind-boggling technological marvels, theoretical physics is failing to advance. There may be many reasons for this but nothing has been more stupefying than the belief that there is no "real" reality.

Theoretical physics has replaced reality with a house of cards. You know what happens when you pull one card out- it all collapses.

Remember the spoons? Suppose you were fooled and accepted the idea that the magician bent the spoon with his mind.

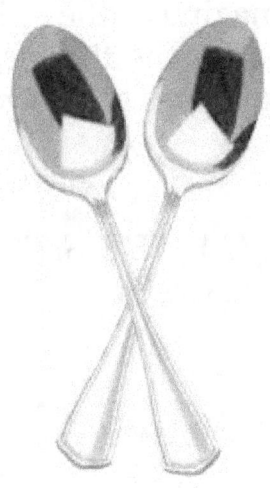

Then you ask scientists to explain how this mental energy works. Theories emerge. The more plausible theories become popular. Millions of books are written and sold.

Several scientists become famous for their variations of these theories. All the universities begin offering courses that teach and compare these theories.

Millions of college students become versed in these competing theories.

Then the magician comes forward and reveals that he just bent the other spoon, not with his mind, but with his thumb.

Can you imagine the humiliation and embarrassment? Maybe the picture of a house of cards collapsing flashes across your mind.

THAT IS EXACTLY THE SITUATION WE NOW HAVE ON OUR HANDS.

"Why Theory" will announce to the world that there is no wave/particle duality and no observer created reality.

We misunderstood a magic trick, the double slit experiment. I know this is going to be painful to admit, but we have been wrong for 100 years. That's okay. We will be right for the next 100 years!

Why Theory

<u>Special Note to Albert Einstein:</u>

We have done it. We have replaced the

"Observer Created Reality" with

"A MODEL OF REALITY"